Satellite Broadcasting Fundamentals

By

Md Tapon Mahamud Jony
B.sc Eng (BD), MBA (UK)

Abstract

You will learn full concept of the term live broadcast technology. From our interest we collected approx all information about live broadcast technology. Here combines an analytical approach to standard of live broadcast technology according the country leading two full HD automated television channel. You will get the information about live broadcast satellite communication including others live broadcast technique ie-live broadcasting with fiber optic, live via DSNG/SNG, live via OB. This book also presented new live streaming technology which created a new era in world live technology and gives a better live solution to especially social media platforms. You will able to learn the basics of modulation-demodulation technique, orbital, transponder, LNB, Encoder etc.

Table of content

Chapter 1

Introduction of satellite communication　　　　　　　　　Page No

1.1　What is satellite?　　　　　　　　　　　　　　　　　　9

1.2　Brief of satellite communication system　　　　　　　　10

1.3　The ground station　　　　　　　　　　　　　　　　　11

Chapter 2

Major parts of satellite

2.1　Major parts of satellite　　　　　　　　　　　　　　　　12

2.2.　Footprint ,Transponder　　　　　　　　　　　　　　　13

2.3　Satellite Orbital's (LEO, MEO, GEO)　　　　　　　　　14

Chapter 3

Television broadcast standard

3.1　Television Signals　　　　　　　　　　　　　　　　　　16

3.2　Broadcast Satellite Service　　　　　　　　　　　　　　18

3.3　Analog TV broadcast format standard　　　　　　　　　19

3.4　Aspect Ratio　　　　　　　　　　　　　　　　　　　　20

Chapter 4

Live Broadcasting Technology

4.1　Live broadcasting technique　　　　　　　　　　　　　　21

4.2　Live coverage by Optical Fiber　　　　　　　　　　　　21

4.3　Implementation of live via optic fiber　　　　　　　　　22

Chapter 5

Live Broadcasting by DSNG/SNG   Page No

5.1 Transmission over SNG/DSNG   24

5.2 Figure of some SNG/DSNG   25

5.3 DSNG Equipments   26

5.4 Transmission diagram of DSNG   26

5.5 DSNG/SNG Modulation Type   27

5.6 DSNG audio & video conversion   29

5.7 DSNG antenna type with figure   29

Chapter 6

Live Broadcasting by OB

6.1 Live broadcasting by OB van   31

6.2 Working procedure of OB Van   32

Chapter 7

Live Broadcasting by Internet Streaming

7.1 Internet streaming live technology   35

Chapter 8

Satellite Uplink & Downlink

8.1 Downlink Frequency   38

8.2 Uplink Frequency   38

8.3 Future Works   39

8.4 References and Citations 40

List of Figure

Fig No	Name of the Figure	Page No
1.1	Satellite	09
2.1	Satellite Uplink-downlink	12
2.2	Major parts of satellite	12
2.3	Foot Print	13
2.4	APSTAR 2R satellite foot print	13
2.5	Satellite Transponder	14
2.6	Satellite orbital comp. table	15
3.1	Frequency Spectrum Curve	16
4.1	Live transmission via optical fiber	21
5.1	Diganta Television DSNG	25
5.2	BBC DSNG	25
5.3	DSNG Equipment's	27
5.4	Basic GPS System	27
5.5	Transmission over DSNG	28
5.6	DSNG Antenna Figure	30
6.1	Internal working activity of an OB van	31
6.2	Online switcher or Video Mixer	32
6.3	Audio Mixer	32
6.4	Vizrt GFX Editing terminal	33
6.5	Video Tape Recorder	33
6.6	Program Director of PCR	33
7.1	Adobe Media Server	35
7.2	Adobe Media Encoder terminal	35
7.3	Controlling Terminal	36
7.4	Adobe Live streaming console	37

List of Table

No.	Name of Table	Page No
2.1	Satellite Orbital comparison table	15
5.1	QPSK phase encoding table	28
8.1	Table of Downlink	38
8.2	Table of Uplink	38

Live Broadcast Technology

Chapter 1

What is satellite?

1.1 Satellite is an-

An earth-orbiting device used for receiving and transmitting signals. Each satellite has a number of transponders which receive the signal and bounce it back to earth; Satellite is a microwave repeater in the space. There are about 750 satellites in the space; most of them are used for communication.

TELSTAR 10 (APSTAR 2R)

Associated Press

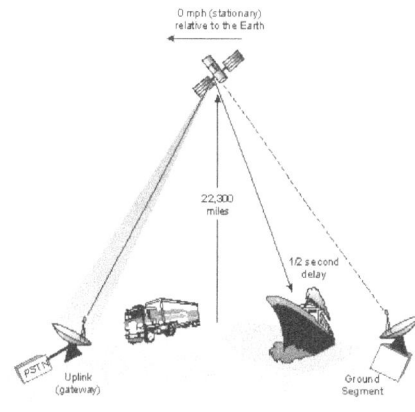

Figure 1.1: Satellite System, Source. Google, Wiki

1.2 Brief of satellite systems:

The basic types of satellite systems include geostationary (GEO), Low Earth Orbit (LEO), Medium Earth Orbit (MEO), and Highly Elliptical Orbit (HEO) satellites. There are also public and private satellite systems such as Television Receive Only (TVRO), Direct Broadcast Satellite (DBS), Global Positioning System (GPS), and multibeam satellite

operations. Geosynchronous satellites orbit the Earth on repeatedly regular points over time. Each GEO satellite is stationary over one spot above the equator and therefore does not need any tracking from receiving and transmitting antennas on the Earth. GEO satellites enable the coverage of weather events. They are especially useful for monitoring severe local storms and tropical cyclones. They are best for television transmission and high-speed data transmission.

Low Earth Orbit (LEO) satellite systems fly very closely to the surface of the Earth, up to 1,500 kilometers in altitude. They deliver more significant voice quality over GEOs and transmit signals with a small margin of delay. Some LEO systems are designed for satellite phones or global mobile personal communications systems. These can carry voice traffic among other data formats.

Medium Earth Orbit (MEO) satellite systems operate at about 10,000 kilometers above the Earth, making it lower than GEO orbits but higher than LEO orbits. They have a larger capacity than LEOs. This enables them more flexibility in satisfying shifting market demands for voice or data services.

Highly-elliptical orbit (HEO) satellite systems orbit the Earth in an elliptical path unlike the LEO's and GEO's circular paths. Its elliptical orbit allows a wider view of the Earth and maximizes the amount of time each satellite spends in viewing populated areas. It therefore requires fewer satellites than LEOs while providing an excellent line of sight.

TVRO (Television Receive-Only) and DBS (Direct Broadcast Satellite) are satellite TV systems. TVRO relies on unencrypted feeds transmitted using open standards. They are also often referred to as C-Band Satellite TV, Big Dish TV, or Big Ugly Dish (BUD).

DBS works on higher frequencies. It is capable of transmitting higher power signals. DBS was primarily intended for home reception. This is why it is also known as Direct to Home satellite.

DBS satellites are owned by satellite TV providers. This means it is restricted to provide free channels.

A global positioning satellite system receives and compares the signals from orbiting GPS satellites to determine geographic location. Each satellite can transmit its exact location with a timed reference signal which the GPS uses to determine the distance between satellites. The location can be marked by calculating the point at which all distances cross. The information can be displayed in latitude or longitude format, or as a position on a computer map.

The multibeam satellite operation uses Spatial Division Multiple Access (SDMA) technology. This allows a single satellite to simultaneously communicate to 2 different satellites using several directional antennas.

The major types of satellite dishes are motor-driven dishes, multi-satellites, VSAT, and ad hoc satellites. Other types include DTH, SMATV, CABD, automatic tracking satellite dishes, and big ugly dishes.

1.3 The Ground Station:

This is the earth segment. The ground station's job is two-fold. In the case of an uplink, or transmitting station, terrestrial data in the form of baseband signals, is passed through a baseband processor, an up converter, a high powered amplifier, and through a parabolic dish antenna up to an orbiting satellite. In the case of a downlink, or receiving station, works in the reverse fashion as the uplink, ultimately converting signals received through the parabolic antenna to base band signal.

Chapter 2

2.1 Major parts of satellite:

- Space Station (Satellite)
- Earth Station

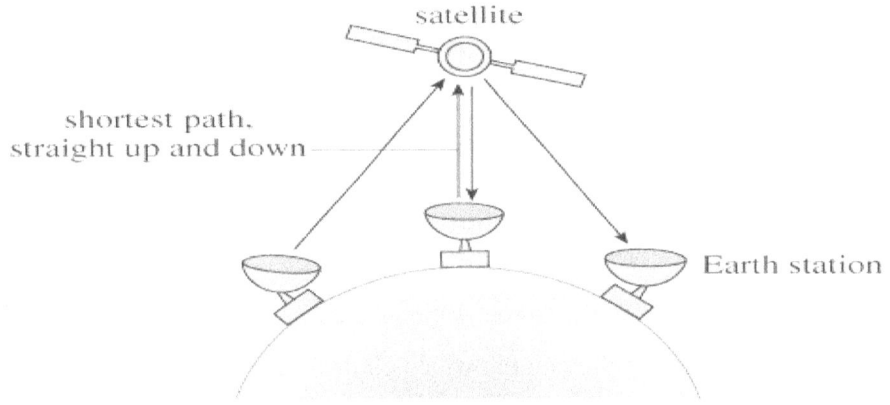

Figure 2.1 : Uplink & Downlink, source Google

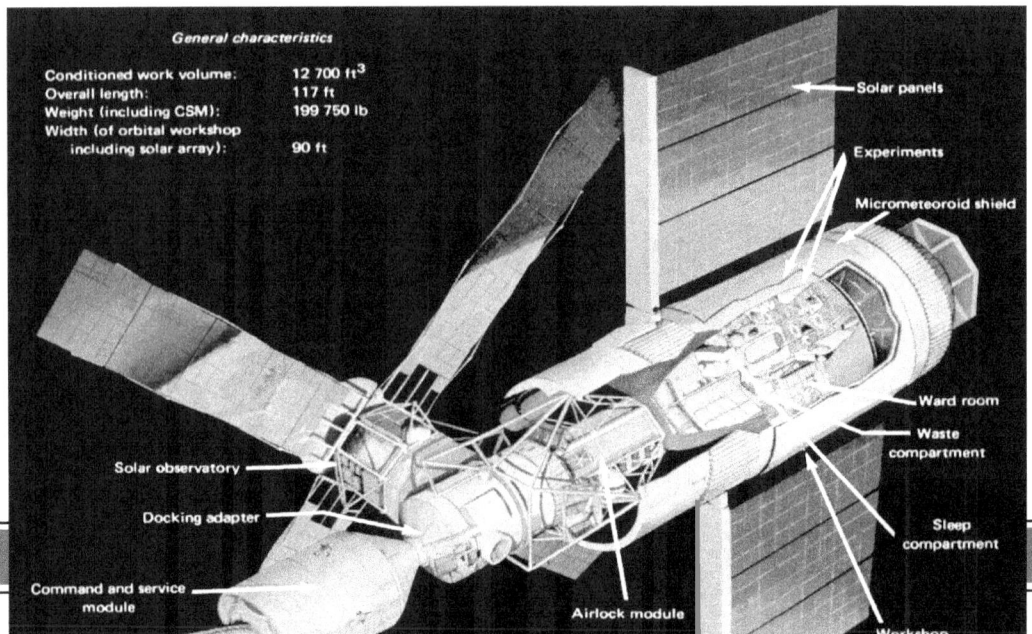

Figure 2.2: Major parts of satellite, source Google

2.2.1 Foot print:

The geographical range of reception of a satellite signal, usually given with signal strength.

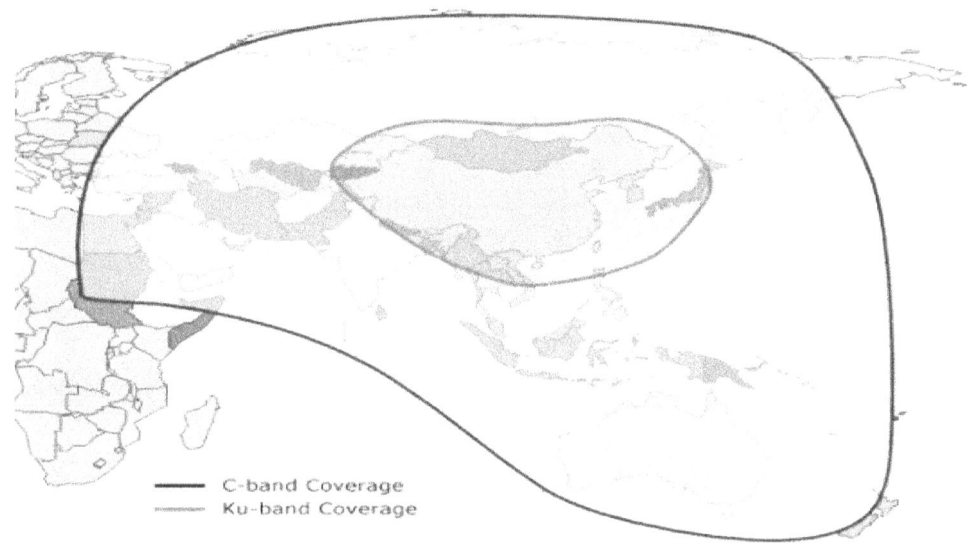

Figure 2.3: Foot print of a satellite, source Wiki

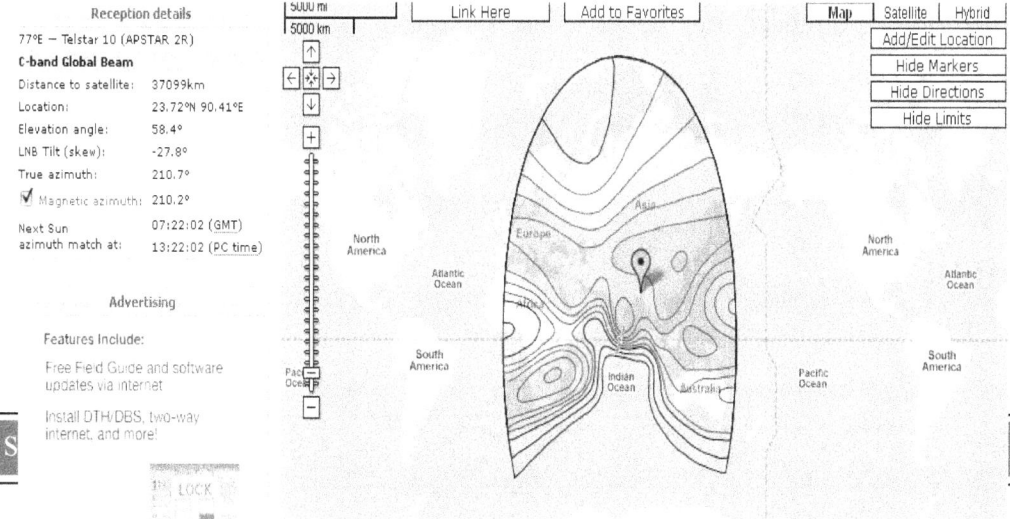

Figure 2.4: APSTAR 2R satellite foot print, Source Wiki

2.2.2 Transponder:

The device of a communications satellite that receives the uplink signal sent from the ground, shifts its frequency to the downlink frequency, amplifies it, and transmits it to the ground.

Figure 2.5: Satellite Transponder, Source Wiki

2.3 Satellite Orbital's:

2.3.1 LEO:

Low Earth Orbit (LEO) satellite systems fly very closely to the surface of the Earth, up to 1,500 kilometers in altitude. They deliver more significant voice quality over GEOs and transmit signals with a small margin of delay. Some LEO systems are designed for satellite phones or global mobile personal communications systems. These can carry voice traffic among other data formats.

2.3.2 MEO:

Medium Earth Orbit (MEO) satellite systems operate at about 10,000 kilometers above the Earth, making it lower than GEO orbits but higher than LEO orbits. They have a larger capacity than LEOs. This enables them more flexibility in satisfying shifting market demands for voice or data services.

2.3.3 GEO:

GEO stands for Geostationary Earth Orbit. This refers to satellites that are placed in orbit such that they remain stationary relative to a fixed spot on earth. If a satellite is placed at 35,900 km above the earth, its angular velocity is equal to that of the earth, thereby causing it

to appear to be over the same point on earth. This allows for them to provide constant coverage of the area and eliminate blackout periods of ordinary orbiting satellites, which is good for providing television broadcasting. However their high altitude causes a long delay, so two way communications, which would need to be uploaded and then downloaded over a distance of 72,000 km, are not often used with this type of orbit.

Satellite orbital comparison:

Type	LEO	MEO	GEO
Description	Low Earth Orbit	Medium Earth Orbit	Geostationary Earth Orbit
Height	500-1000 miles	6000-12000 miles	22,282 miles
Time in LOS	15 min	2-4 hrs	24 hrs
Merits	Lower launch costs. Short round trip signal delay. Small path loss.	Moderate launch cost. Small round trip delays.	Covers as much as 42.2% of the earth's surface. Ease of tracking. No problems due to doppler.
Demerits	Shorter life, 5-8 years. Encounters radiation belts.	Larger delays. Greater path loss than LEO's.	Very large round trip delays. Expensive Earth Stations due to weak signals.

Table 2.1: Satellite Orbital comparison table, Source Internet

Chapter 3

Television Signals

Satellites have been used for since the 1960's to transmit broadcast television signals between the network hubs of television companies and their network affiliates. In some cases, an entire series of programming is transmitted at once and recorded at the affiliate, with each segment then being broadcast at appropriate times to the local viewing populace. In the 1970's, it became possible for private individuals to download the same signal that the networks and cable companies were transmitting, using c-band reception dishes. This free viewing of corporate content by individuals led to scrambling and subsequent resale of the descrambling codes to individual customers, which started the direct-to-home industry. The direct-to-home industry has gathered even greater momentum since the introduction of digital direct broadcast service.

We are very familiar with the C and Ku bands. These bands are normally used for digital TV transmission. Some of our readers also know the S band. However, the frequency spectrum that can be used for satellite communication is not limited to the above mentioned ones. Before we move on to Ka-Band, let's have a look first at the whole radio frequency spectrum

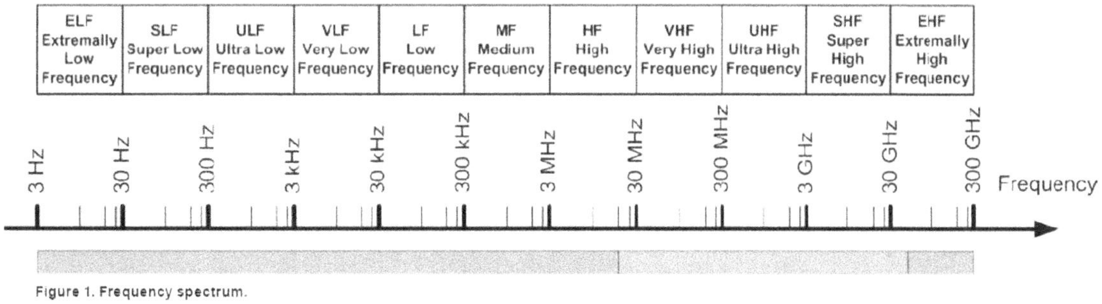

Fig: 3.1: Frequency Spectrum curve, Source Wiki

3.1 Television Signals

3.1.1 C-band

C-Band (3.7 - 4.2 GHz) - Satellites operating in this band can be spaced as close as two degrees apart in space, and normally carry 24 transponders operating at 10 to 17 watts each. Typical receive antennas are 6 to 7.5 feet in diameter. More than 250 channels of video and 75 audio services are available today from more than 20 C-Band satellites over North America. Virtually every cable programming service is delivered via C-Band. The communications C-band was the first frequency band that was allocated for commercial telecommunications via satellites. The same frequencies were already in use for terrestrial microwave radio relay chains. Nearly all C-band communication satellites use the band of frequencies from 3.7 to 4.2 GHz for their downlinks, and the band of frequencies from 5.925 GHz to 6.425 GHz for their uplinks. Note that by using the band from 3.7 to 4.0 GHz, this C-band overlaps somewhat into the IEEE S-band for radars.

The C-band communication satellites typically have 24 radio transponders spaced 20 MHz apart, but with the adjacent transponders on opposite polarizations. Hence, the transponders on the same polarization are always 40 MHz apart. Of this 40 MHz, each transponder utilizes about 36 MHz. (The unused 4.0 MHz between the pairs of transponders acts as "guard bands" for the likely case of imperfections in the microwave electronics.)

3.1.2 Ku-Band

Ku Band (11.7 - 12.2 GHz) - Satellites operating in this band can be spaced as closely as two degrees apart in space, and carry from 12 to 24 transponders that operate at a wide range of powers from 20 to 120 watts each. Typical receive antennas are three to six feet in diameter. More than 20 FSS Ku-Band satellites are in operation over North America today, including several "hybrid" satellites which carry both C-Band and Ku-Band transponders. Prime Star currently operates off Satcom K-2, an FSS or so-called "medium-power" Ku-Band satellite. Alpha Star also uses an FSS-Ku Band satellite,
Telestar 402-R.

3.2 Broadcasting Satellite Service (BSS)

Ku-Band (12.2 - 12.7 GHz) - Satellites operating in this band are spaced nine degrees apart in space, and normally carry 16 transponders that operate at powers in excess of 100 watts. Typical receive antennas are 18 inches in diameter. The United States has been allocated eight BSS orbital positions, of which three (101, 110 and 119 degrees) are the so-called prime "CONUS" slots from which a DBS provider can service the entire 48 contiguous states with one satellite. A total of 32 DBS "channels" are available at each orbital position, which allows for delivery of some 250 video signals when digital compression technology is employed.

3.2.1 DBS

DBS (Direct Broadcast Satellite) -The transmission of audio and video signals via satellite direct to the end user. More than four million households in the United States enjoy C-Band

DBS. Medium-power Ku-Band DBS surfaced in the late 1990s with high power Ku-Band DBS.

Analog TV broadcast format standard

1. NTSC (National Television Standard Corporation)

2. PAL (Phase Alternating Line)

3. SECAM (Sequential cooler a memoires)

3.2.2 : NTSC:

NTSC receivers have a tint control to perform color correction manually. If this is not adjusted correctly, the colors may be faulty. The PAL standard automatically cancels hue errors by phase reversal, so a tint control is unnecessary. Chrominance phase errors in the PAL system are cancelled out using a 1H delay line resulting in lower saturation, which is much less noticeable to the eye than NTSC hue errors.

3.2.3 : PAL:

PAL, short for Phase Alternating Line, is an analogue television color encoding system used in broadcast television systems in many countries. Other common analogue television systems are NTSC and SECAM. The articles on broadcast television systems and analogue television further describe frame rates, image resolution and audio modulation. For discussion of the 625-line / 50 field (25 frame) per second television standard

3.2.4 : SECAM:

SECAM, also written SECAM (Sequential cooler a memoires, French for "Sequential Color with Memory"), is an analog color television system first used in France. A team led by Henri de France working at Companies Francoise de Television (later bought by Thomson, now Technicolor) invented SECAM. It is, historically, the first European color television standard.1H delay line to allow decoding of only the odd or even lines. For example, the chrominance on odd lines would be switched directly through to the decoder and also be stored in the delay line. Then, on even lines, the stored odd line would be decoded again. This method effectively converted PAL to NTSC. Such systems suffered hue errors and other problems inherent in NTSC and required the addition of a manual hue control.

Aspect Ratio

3.3 Aspect Ratio:

The aspect ratio of an image is the ratio of the width of the image to its height, expressed as two numbers separated by a colon. That is, for an $x:y$ aspect ratio, no matter how big or small the image is, if the width is divided into x units of equal length and the height is measured using this same length unit, the height will be measured to be y units. For example, consider a group of images, all with an aspect ratio of 16:9. One image is 16 inches wide and 9 inches high. Another image is 16 centimeters wide and 9 centimeters high. A third is 8 yards wide and 4.5 yards high. Aspect ratios are mathematically expressed as $x:y$ (pronounced "x-to-y") and $x \times y$ (pronounced "x-by-y"), with the latter particularly used for pixel dimensions, such as 640×480. Cinematographic aspect ratios are usually denoted as a (rounded) decimal multiple of width vs unit height, while photographic and video graphic aspect ratios are usually defined and denoted by whole number ratios of width to height. In digital images there is a subtle distinction between the Display Aspect Ratio (the image as displayed) and the *Storage Aspect Ratio* (the ratio of pixel dimensions)

An aspect ratio is the ratio between the width and height of a film image. TV - 4:3 Computer Monitor – 16:9

3.3.1 SDTV:

Standard-definition television refers to television systems that have a resolution that meets standards. The term is usually used in reference to digital television.

640x480 or lower Aspect Ratio: 4:3

3.3.2 HDTV:

A broadcast technology in which the number of *scan lines* of the video image is increased and the size of the *pixels* decreased.

1280×720 or higher Aspect Ratio: 16:9

Chapter 04

Live Broadcast Technique

- ❖ Live Broadcast by Fiber Optic
- ❖ Live Broadcast by SNG/DSNG
- ❖ Live Broadcast by OB Van
- ❖ Internet Live Streaming.

Live transmission with Fiber Optic

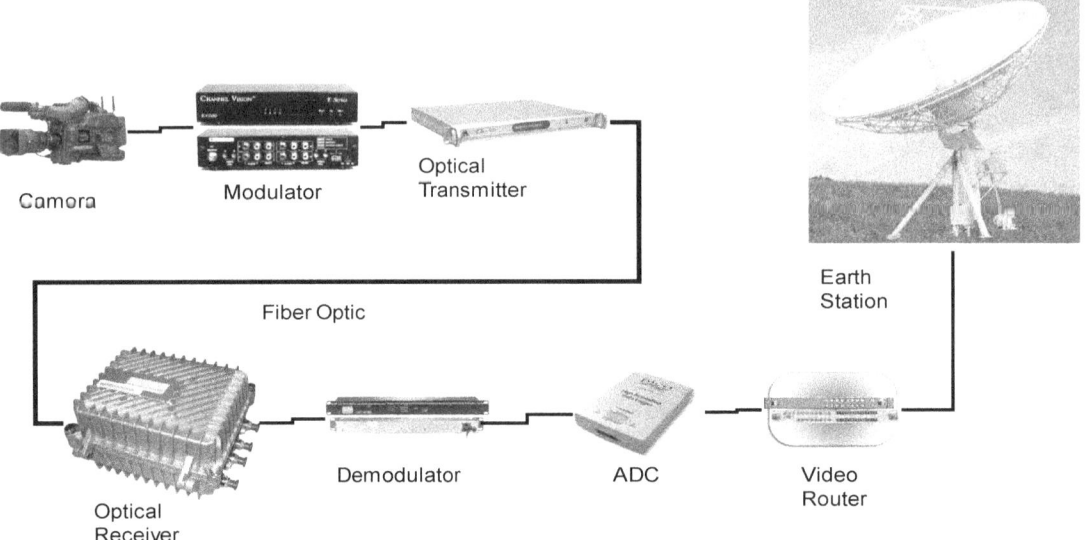

Figure 4.1: Live transmission via optical fiber.

Implementation process of Live via optical fiber Transmission:

- **Camera:**

Shot the news/program then transfer the DV25 or raw video files to modulator via fire wire cable or usb cable.

- **Modulator:**

Modulation is the addition of information (or the signal) to an electronic or optical signal carrier. Modulation can be applied to direct current (mainly by turning it on and off), to alternating current, and to optical signals. One can think of blanket waving as a form of modulation used in smoke signal transmission (the carrier being a steady stream of smoke). Morse code, invented for telegraphy and still used in amateur radio, uses a binary (two-state) digital code similar to the code used by modern computers. For most of radio and telecommunication today, the carrier is alternating current (AC) in a given range of frequencies.

Receive the video signal from camera then modulate the signal and transfer to the optical transmitter.

- **Optical Transmitter:**

A device that accepts an electrical signal as its input, processes this signal, and uses it to modulate an opto-electronic device, such as an LED or an injection laser diode, to produce an optical signal capable of being transmitted via an optical transmission medium

Optical Transmitter receive the signal from modulator, then encrypt the signal and transfer to the Optical Receiver via optical cable

Receiving

- **Optical Receiver:**

A device that detects an optical signal, converts it to an electrical signal, and processes the electrical signal

Receive the modulated signal from optical fiber, then after decrypt transfer the signal to demodulator

- **Demodulator:**

Receive the modulated signal then demodulated the signal and transfer it to ADC to get digital signal.

- **ADC**:

  Analog to digital Converter receive the demodulated signal and convert it to digital signal and finally transfer to the video router.

- **Video Router**:

  Routing switchers are basically an evolutionary line of products that can be traced back at least as far as patch panels. For about two decades, little changed other than rack density, signal quality specifications and warranty offerings. However, in the last 10 years, much has changed in keeping with the function that routing performs in the modern facility and the evolution of signal types.

  Receive the digital signal from ADC, finally transmit the signal to satellite.

Chapter 05

Live telecasting with SNG/DSNG

What is SNG/DSNG?

5.1.1 The SNG:

Satellite News Gathering is a technology of live coverage system, where modulator, encoder, Up converter is individual used to transmit a signal to satellite. Antenna have to set manually. Satellite news gathering (SNG) is the use of mobile communications equipment for the purpose of worldwide news casting.

5.1.2 The DSNG:

Digital Satellite news Gathering is a technology of live broadcasting system, where the modulator, encoder, up convert is built-in with a complete tools known as Exciter. That's why It's called Digital SNG. Here antenna can be set automatically. Then transfer the modulated signal to Satellite. A modern DSNG van is a sophisticated affair, capable of deployment practically anywhere in the civilized world. Signals are beamed between a geostationary satellite and the van, and between the satellite and a control room run by a broadcast station or network. In the most advanced systems, Internet Protocol (IP) is used. Broadcast engineers are currently working on designs for remotely controlled, robotic DSNG vehicles that can be teleoperated in hostile environments such as battle zones, deep space missions, and undersea explorations without endangering the lives of human operators.

5.2.1 Figure of SNG/DSNG:

Image credit, Diganta Television'2011

5.2.2: DSNG:

Image source Google

DSNG Equipments

5.3 Transmission Diagram of DSNG:

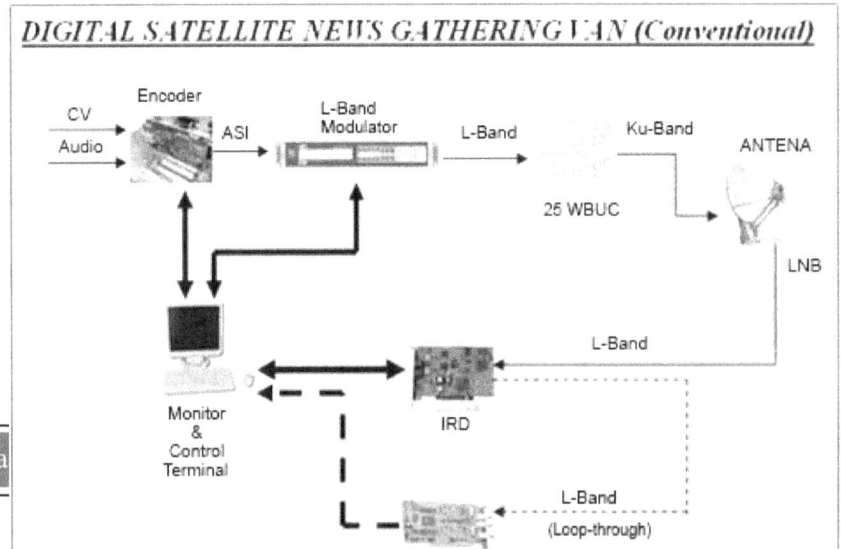

Figure 5.3: DSNG Equipment's, Source. Wiki

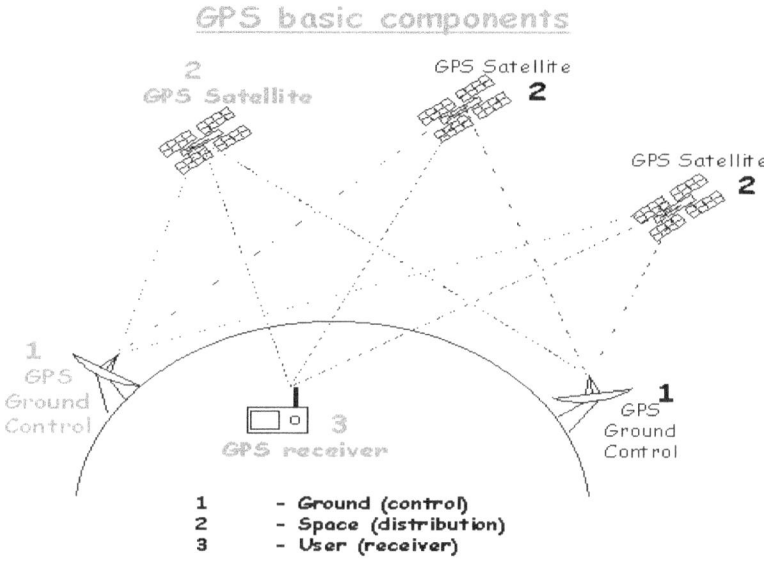

Figure 5.4: Basic GPS systems, Source Google

5.4 Transmission over DSNG/SNG:

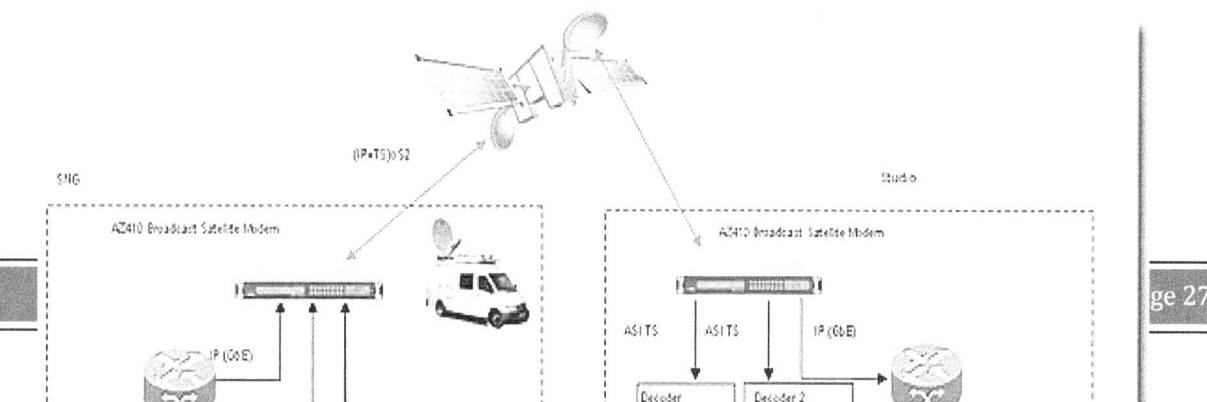

Fig 5.5: Transmission over DSNG, Source SBE

5.5 DSNG/SNG Modulation Type:

5.5.1 Modulation:

Generally in DSNG/SNG the modulation type is QPSK for MPEG2 or MPEG4. It various to different with L-QPSK, H-QPSK.

QPSK (Quadrature Phase Shift Keying) is a phase modulation algorithm. Phase modulation is a version of frequency modulation where the phase of the carrier wave is modulated to encode bits of digital information in each phase change. The "PSK" in QPSK refers to the use of Phased Shift Keying. Phased Shift Keying is a form of phase modulation which is accomplished by the use of a discrete number of states. QPSK refers to PSK with 4 states. With half that number of states, you will have BPSK (Binary Phased Shift Keying). With twice the number of states as QPSK, you will have 8PSK. The "Quad" in QPSK refers to four phases in which a carrier is sent in QPSK: 45, 135, 225, and 315 degrees.

5.5.2 QPSK Encoding:

Because QPSK has 4 possible states, QPSK is able to encode two bits per symbol.

Phase	Data
45 degrees	Binary 00
135 degrees	Binary 01
225 degrees	Binary 11
315 degrees	Binary 10

Table 5.1: QPSK Phase, Source Internet

5.6 Video & Audio conversion in DSNG/SNG

Video and audio receive by a compact package from camera. Earlier there used individual format of audio and video. But now a day's video came through the same source. That's why now audio and video might be seconds of displacement. To eliminate this problem we use a masking and de masking equipment of various brands. Audio & video displacement problem only occurred while compressing in MPEG4 format, but most of the TV channel of Bangladesh use MPEG2, that's why there didn't need masking equipment.

5.7 SNG/DSNG/OB Antenna Type: (comparison between 3 tv channel).

- **Independent** Tv use:
  Station main antenna : 4.5 Meter
  DSNG antenna : 1.8 Meter

- **Diganta** Tv use:
  Station main antenna : 4.5 Meter
  SNG Antenna : 2.2 Meter

- **Somoy** Tv use:
  Station main antenna : 4.5 Meter
  SNG Antenna : 2.4 Meter

Data Source. Wiki Bangladeshi Satellite TV

DSNG Antenna Types

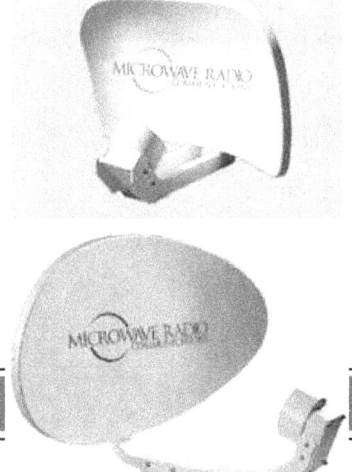

Figure 5.6: DSNG Antenna Figure, Image source Google

Chapter 6

Live broadcasting by OB (Outdoor Broadcasting Van)

6.1 What is OB?

While we need to do live from outdoor then OB van is ideal equipments for a TV station. We can called it a mini television station, where there is a vision mixer, audio controller, editing panel, GFX updater, VTR etc. We can send the RF signal directly to satellite.

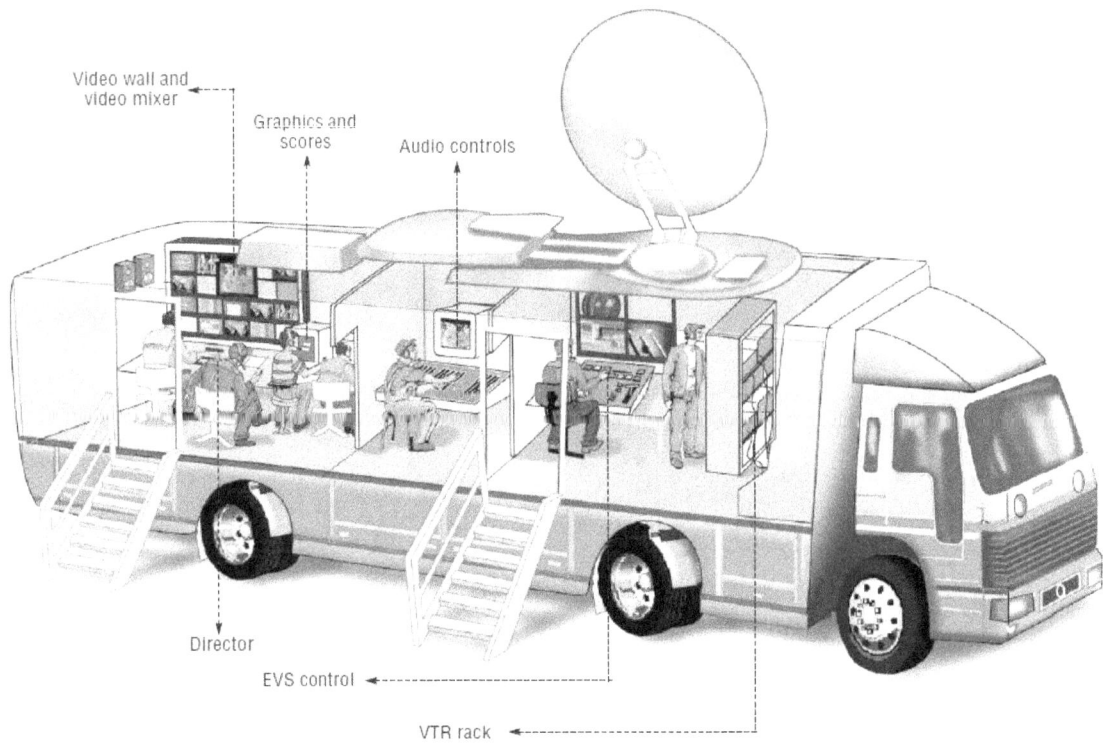

Figure 6.1: Internal working activity of an OB van, Source OBS

Working procedure of OB van

6.2.1 Video Mixer:

The video mixer is one type of online switcher machine. The operator took decision from which camera it will be on-air. He can also give video effects from online switcher console.

Figure 6.2: Online switcher or Video Mixer

6.2.2 Audio Mixer:

Audio Mixer operator balances the audio from camera. Because during recording with camera there are so many interferences he eliminates the interference from source and gives a better output through his audio mixer.

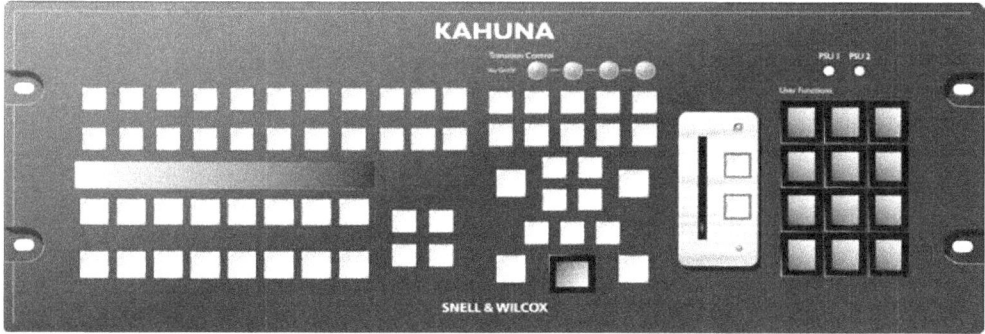

Figure 6.3: Audio Mixer

6.2.3 GFX Editor:

Graphics editor can edit the still or motion graphics. Also during live games like cricket, football, volleyball he can give score through gfx card via vizrt systems.

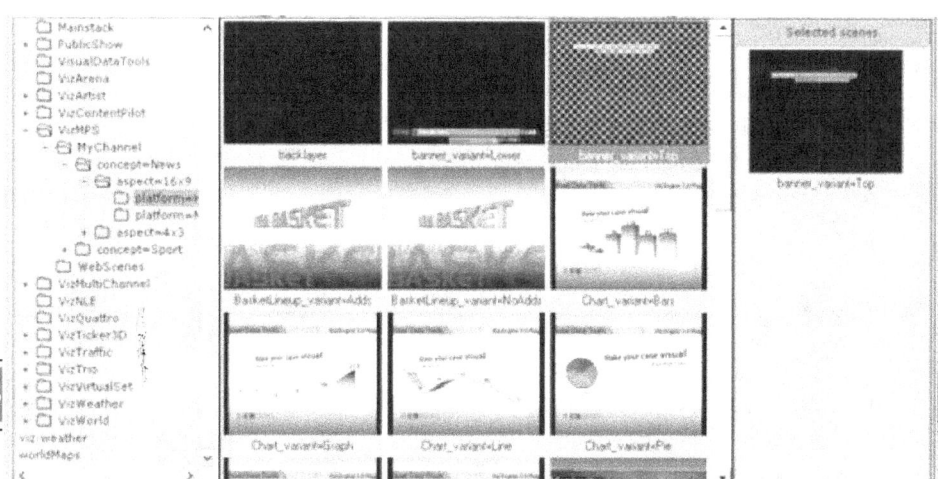

Figure 6.4: Vizrt GFX Editing terminal

6.2.4 VTR:

Video Tape Recorder is one kind of recorder. After recording at camera VTR operator insert the disc/tape to VTR then capture the video and deliver it to server.

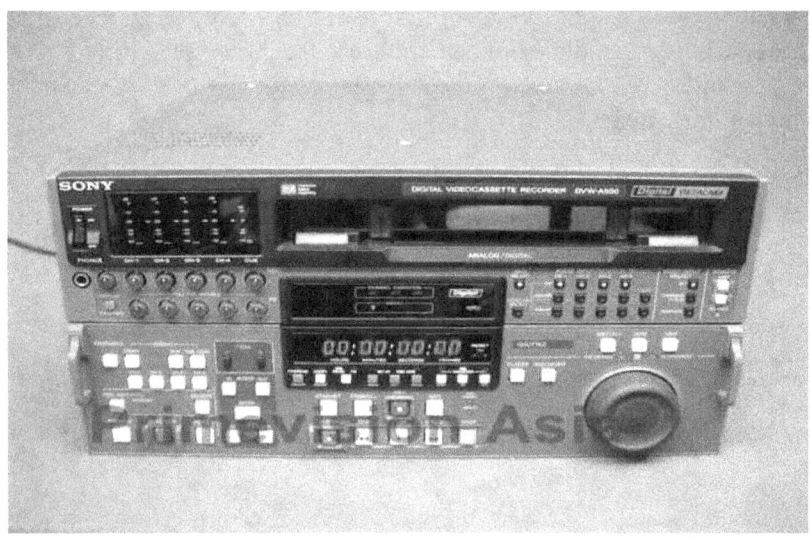

Figure 6.5: Video Tape Recorder, Source Primevision Asia

6.2.5 : Director:

Director took the whole decision of the live program. With which cam it will be on-air. Any change during live program. Give direction to the artist. All these decisions must come through the director.

Figure 6.6: Program Director of PCR, Source Internet

Chapter 7

Internet streaming live technology

7.1 :Live Streaming Technology:

World becomes closer day by day. Now we can live any video by using camera, and telecast it via internet with using some tools or application. Adobe flash media encoder is such an application for live streaming. Here we will describe shortly about this application.

Adobe Flash Media Live Encoder live audio and video capture software is a media encoder that streams audio and video in real time to Flash Media Server software or Flash Video Streaming Service (FVSS).

- Sporting events
- Concerts
- Webcasts
- News
- Educational events

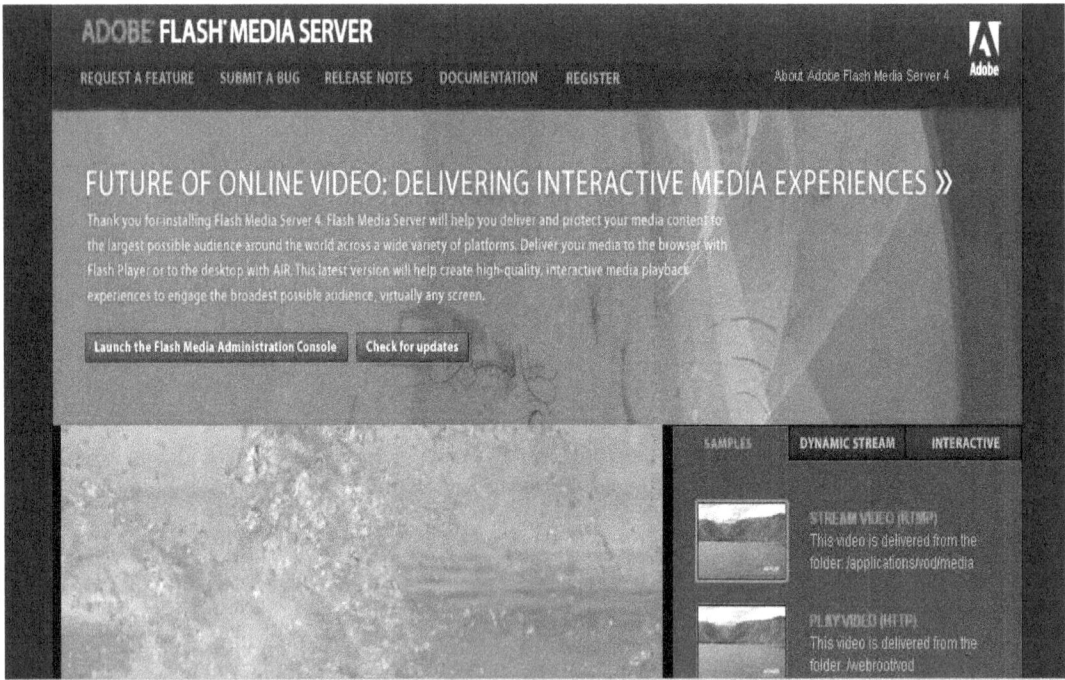

Figure 7.1: Adobe Media Server , Source Adobe website

7.1.1 Working Systems?

❖ Need to install Flash Media Server first of all at server computer
❖ Select the video format with necessary configuration as TV channel required.
❖ Install the client end Flash Media Encoder software.
❖ Record video and transfer to the server through internet
❖ Take the streaming video into a source and playback the video to destination channel.

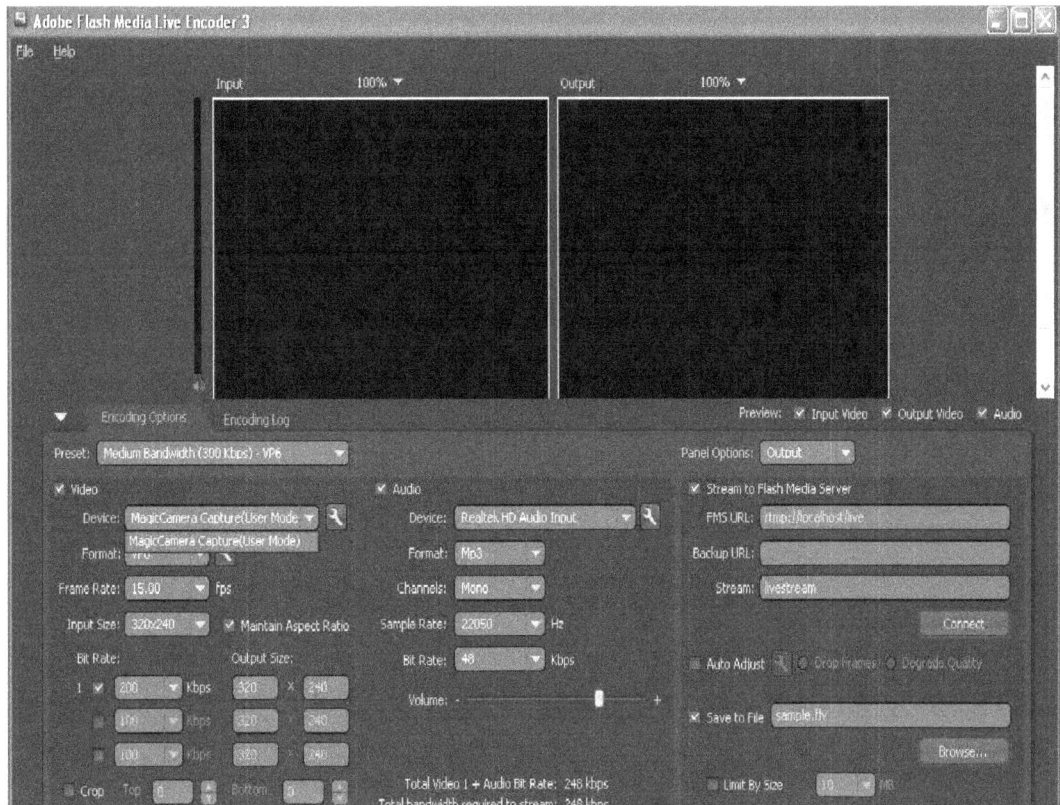

Figure 7.2: Adobe Media Encoder terminal, Source ITV

Controlling Interface:

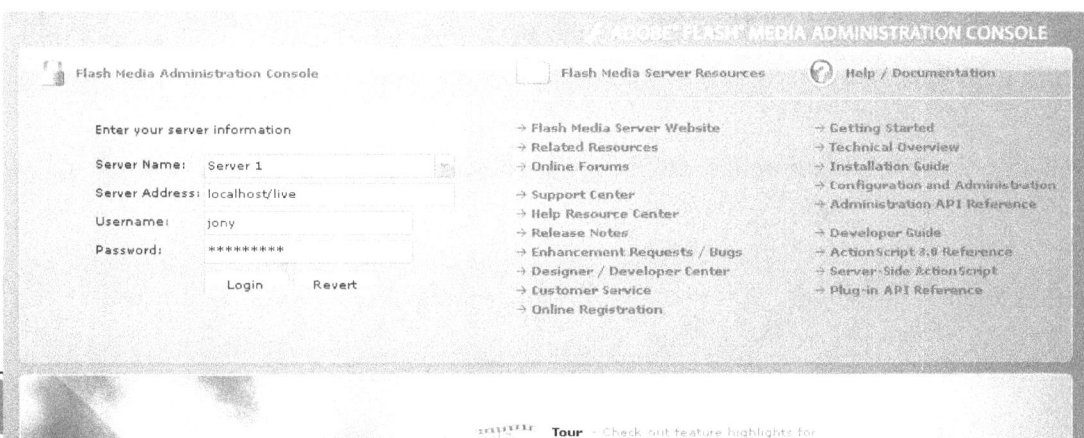

Figure 7.3: Controlling terminal, Source ITV

Media Live streaming console:

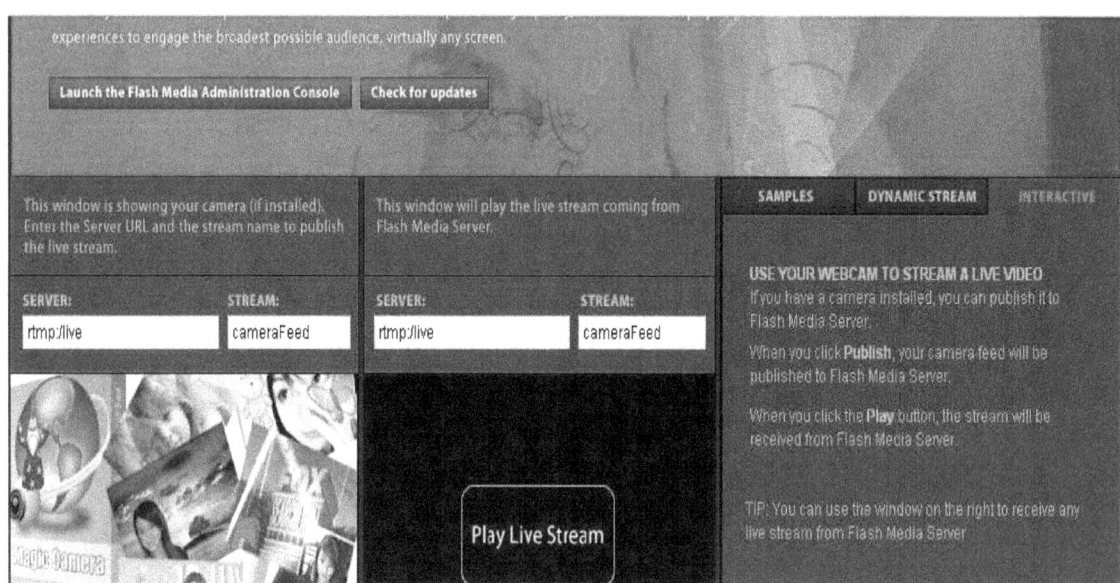

Figure 7.4: Media Live streaming console, Source ITV

Chapter 8

Uplink & Downlink Frequency

8.1 Downlink:

Downlink is the signal path from the satellite toward the earth. In Bangladesh we use only C band frequency which range is 3.7 to 6.42 GHz

Satellite Band	Downlink Frequency
L Band	0.5 - 1.5 GHz
C Band	3.7 - 4.2 GHz
Ku Band	11.7 - 12.7 GHz

Table 8.1: Downlink Frequency, Source Wiki

8.2 Uplink:

Uplink is the signal path from an earth station to a satellite

Satellite Band	Uplink Frequency
L Band	1.55 – 3.5 GHz
C Band	5.925 - 6.425 GHz
Ku Band	14 - 14.5 GHz
Ka Band	27.5 - 31 GHz

Table 8.2: Uplink Frequency, Source Wiki

8.3 Future Works

Even though various critical aspects of the tactical private Satellite Broadcasting have been demonstrated throughout the course of research and, since the tactical private satellite communication has been developed as a concept to this point; there are areas that are suggested for further study within the context of the tactical private satellite broadcasting. Each of the areas for future research will serve to increase the understanding of the

Operational implementation of this conceptual network, and broaden the overall utility of the system.

The first aspect for future research deals with the actual implementation of the concepts defined with regard to the base station/gateway device. This is to say that commercial satellite service providers may have some mechanism to provide a generic terminal connected to their network the capabilities outlined of the gateway/base station device.

Technology improved daily. This is new today, it will become old tomorrow but fundamentals always remain same. Keep learning and improve skills.

References and Citations

Apstar7 Footprints. (2015, July 12th). Retrieved May 2nd, 2016, from satbeams: https://www.satbeams.com/footprints?norad=38107

Britannica. (2015). Retrieved Nov 8th, 2018, from Television Transmission & Reception: https://www.britannica.com/technology/television-technology/The-television-channel

Cavell, G. C. (2014). *Engineering Handbook*. New York: Routeledge.

Footprint of Satellite. (2017). Retrieved 2017, from Wikipedia: https://en.wikipedia.org/wiki/Footprint_(satellite)

Friedländer, M. B. (2016, Jan 1st). *Streamer Motives and User-Generated Content*. Retrieved Oct 2nd, 2016, from Streamer Motives: https://www.researchgate.net/publication/320137414_Streamer_Motives_and_User-Generated_Content_on_Social_Live-Streaming_Services

Haskell, B. G., Puri, A., & Netravali, A. N. (1997). *Digital Video: An introduction to MPEG2*. New York: International Thomson Publishing.

Higgins, J. (2007). *Satellite News Gathering*. USA: Focal Press.

Jackson, K. G., & Townsend, G. B. (1991). *TV & Video Engineers Referrence book*. Oxford: Butterworth-Heinemann.

Outside Broadcasting. (n.d.). Retrieved from Durban University of Technology: https://www.dut.ac.za/faculty/arts_and_design/video_technology/ob_van/

Weynan, D., Piccin, V., & Weise, M. (2016). *How video Works*. New York: Focal Press.

What to Know About Live Video, Social Media's Latest Craze. (n.d.). Retrieved Nov 3rd, 2016, from Wall Street Journal: https://www.wsj.com/articles/live-video-what-to-know-about-social-medias-latest-craze-1461691861

What's the Difference Between LEO, MEO and GEO. (2016, July 01). Retrieved from Cosmo BC: http://astroblog.cosmobc.com/2019/06/01/difference-leo-meo-geo-satellites/

Wikipedia. (2018, Jul 8th). *What is outside broadcast*. Retrieved Nov 1st, 2018, from Wikipedia: https://en.wikipedia.org/wiki/Outside_broadcasting

Adobe. (2007, May 23rd). Webcasting live video with Flash Media Live Encoder. Retrived Nov 3rd, 2019, from Adobe: https://www.adobe.com/devnet/adobe-media-server/articles/webcasting_fme.html

www.ingramcontent.com/pod-product-compliance
Lightning Source LLC
Chambersburg PA
CBHW081647220526
45468CB00009B/2578